Living with Earth

Naia Jones

Published by the Piscean Press

4930 Waa Street, Suite 102

Honolulu, Hawaii 96821 USA

First Printing: April 2010

ISBN 978-0-578-05699-9

Printed in the United States of America

This book is dedicated to my precious little Ella.

Contents

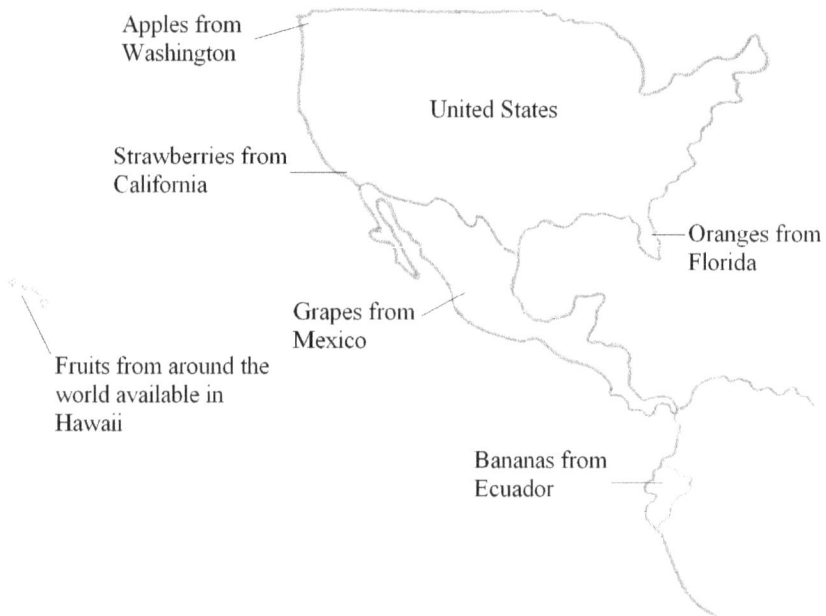

Figure 1: Far Beyond Nomadic Living

INTRODUCTION

Starting with the rise of agriculture, mankind has a long demonstrated history of clever manipulation of the environment for its own means. Well into the 21st century we are far beyond our initial escape from nomadic living. Today we can reside on one continent and eat the fruit of another continent across an ocean on a daily basis!

We continuously create tools that greatly affect our lifestyles such as writing, machines, computers, the internet and the Apple iPod. Ingenuity is surely man's most prized and unique possession. Where is it all leading to? And what will be our next greatest invention? Now that we have conquered our environment our next evolutionary milestone may very well be learning to live in harmony *with* her.

In humanity's youth and exuberance for development and progress, we focused on production irrespective of its consequences on the greater environmental scheme. Like children who don't know any better, we carried on with our activities always relying on the world of abundance with an abandon of care found only in youth. Our so-called development however, has finally caught up with us and in 2010 we are faced with such worldwide environmental challenges as climate change, energy crisis, water scarcity and overpopulation as our world count nears seven billion.

Forced to reckon our youthful abandon with reality, we are in our older years gaining sensitivity to the fundamental truth that we are all connected and that we are responsible for the real consequences our activities have on the environment. Our burgeoning environmental awareness is materializing in nearly all sectors of the marketplace. The auto industry for example, is engaged in a race to carve out a niche for alternatively fueled cars. Toyota's commercials for its third generation Prius represents this new movement well. The Prius drives forward, ushering in the dawning of a new era; flowers bloom, the sun shines, color and life proliferates everywhere in the Prius world of harmony between man, nature and machine.

The product is a telling sign of the times. It is a third generation hybrid electric, mid-size car starting at the competitive price of $22,000. In its third year, the mutant hybrid car is clearly not a one-time gimmick. It is a successful mutation for survival, for a way to consume readily available, clean energy as opposed to depleting, *dirty* energy. Like the mutant hybrid land-sea creatures that evolved toward permanent terrestrial life as a way to seek new habitat, the hybrid electric car is a transition car on its way to fully electric means in order to seek new viable energy sources. We love our machines and going green does not mean we have to give them up. Going green may be just a matter of design.

Figure 2: Toyota Prius

Figure 3: Tiktaalik – Hybrid Land / Sea Creature

As a gage for how far we have come, Walmart infamous for cheap products at the expense of environment and people launched a Sustainability Index and seed-funded a Sustainability Consortium in 2009. The objective of the consortium is to develop the metrics and standards for evaluating sustainable products. There is an obvious shifting in the marketplace due to increased environmental awareness. However, despite an observed monumental shifting to greener products, we are only at the fringes of where we need to be.

As Joel Makower, executive editor of Greener World Media Inc. concluded in 2008, companies are implementing more and more environmental measures, "but moving the needle of environmental progress only slightly, if at all." He summarizes the State of Green Business in an annual report published on GreenBiz.com. The summary at the beginning of 2009 noted a "growing chorus of corporate commitments and actions", just not "in aggregate, addressing planetary problems at sufficient scale and speed."

Just released in February 2010 the latest annual report reveals more of the same; despite the economic downturn and perhaps because of it more and more individual businesses launched green initiatives as a way to cut operating costs and to remain competitive. However as impressive as current efforts are, the sum of it lacks effectiveness for what we are really trying to accomplish.

The reports, while sobering, are no reason for its contributors to be somber. It is an important barometer of where we currently are and where we need to be on our path to environmental harmony. It seems we are on the brink of a new environmental era. If we could only unify and organize our hodge-podge efforts, we could produce effects of global proportions more suitable for addressing the scale and enormity of our environmental challenges.

Thankfully, our development from the Agricultural Age through the Industrial Age and finally to the current Information Age not only provides us with its own challenges, but also gives us some very sophisticated tools to meet them. With advances in telecommunications and transportation we are better equipped more than ever to embrace our opportunities for environmental improvement globally—as one people, as humanity. Together we can amass the strength in numbers required to successfully address our challenges.

Events such as the 2009 Climate Change Summit held in Copenhagen signal recognition of our need to organize together and to collectively find solutions. We are refreshingly stepping in the right direction. But the event also shows how difficult it is to reach consensus across the nations on such large issues without order and conviction. The resulting voluntary political accord, now dubbed the "Copenhagen Accord", is under-whelming and falls short of a mandatory climate treaty to replace the Kyoto Protocol.

So far in our response to crucial global issues we sense the absolute need for unity and collaboration around the world, take small steps in that direction, but lack the overall collective will and organization to execute global solutions at the speed that we need to. It has been said that necessity is the mother of invention. By that rule, we will eventually organize together and implement a new relationship with our environment simply because we have to. The question is, "How far do we enter crisis before we find the resolve to change our ways?"

This book offers insight and the *green prints* to help humanity implement solutions for living with Earth sooner rather than later; to help humanity organize and emerge unified to effectively achieve together far beyond what we ever could apart. Fortunately the guiding principles and the reasons for our monumental paradigm shift are found all around us everywhere we look if we care to see it. Apart from our own conscious evolution the past 200,000 years, Nature has been evolving life on Earth for nearly five billion years. With time on her side, the design found in the natural world is enormously complex and we are no where near discovering all of Nature's secrets much less implementing such complexity in our own designs.

Nevertheless, scientists have given us volumes of research to date and making use of this wisdom is incumbent upon us. A review of Nature's life evolving work reveals three best management practices key to her design success for evolving life on Earth. Implementing these best management practices in our designs will help us move rapidly forward in successfully living with Earth. So without further adieu and with Nature as our guide, let's embark on our knowledge adventure and start from the beginning. And I do mean the beginning.

Part 1

Our Solar System

The birth of our solar system is not exactly well understood since of course no one was around to witness or record the space-shattering event. What follows is our best guess, known as the nebular hypothesis. The nebular hypothesis is based both on proven scientific principles and on observations. Most importantly, it is the story of our celestial origins.

ONE

In the Beginning

A long time ago in a galaxy not so far away...

The solar nebula—a galactic ethereal cloud of cold dust and gas floats along in our very own Milky Way. The solar nebula is gigantic, stretching light years across, its wide expanse owing to a faint pressure of gas. The attractive force of gravity between even the tiniest of mass counteracts the expansive pressure, keeping the molecular cloud bound together in a diffuse, but unified and balanced structure. The whole bulk of it rotates ever so slowly about itself, a remnant effect from the violent explosion that created it.

Zooming out, it is evident that the nebula is part of something even bigger. Not only is the nebula rotating about itself, it rotates along with all other mass in the galaxy, in a giant pinwheel of dust and stars. Spiraling out in all directions, the mass of the entire galaxy remains firmly "pinned" to the same abysmal center of the Milky Way.

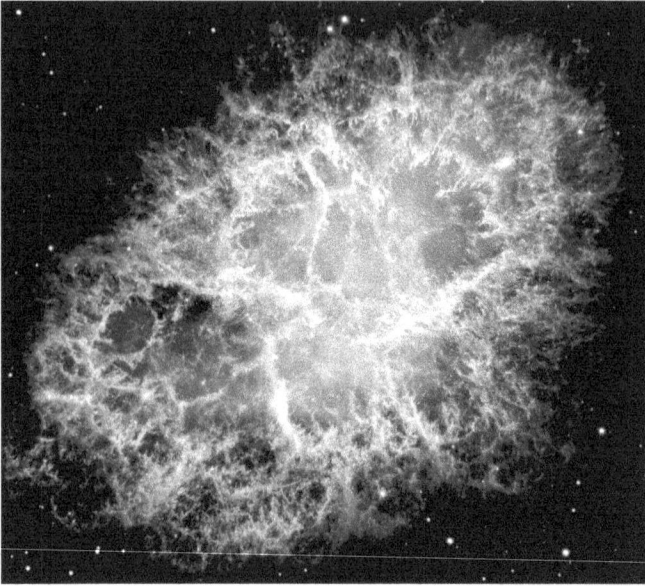

Figure 4: Crab Nebula M1 for Illustration of Nebulae

Nebula within
larger galaxy

Figure 5: Spiral Galaxy

1 light year =   5.9 trillion miles

=   The distance light can travel in one year

All of a sudden a celestial disturbance of some sort, perhaps a shock wave from a neighboring supernova blasts through. The shock causes the solar nebula's delicate structure to collapse under its own weight. The collapse continues in an irreversible chain reaction; as more and more mass collects together its gravity or ability to attract even more mass increases. The collapsing occurs simultaneously in all directions and continues until another balance can be reached.

As the nebula contracts, its slow rotation begins to increase in speed like the ice skater who spins faster as she pulls her arms and legs in, a phenomenon known in physics as the conservation of angular momentum. The conservation of angular momentum derives from the overarching law of the universe—the **law of conservation.** The law of conservation governs all energy and actions in the universe from rotating solar nebulae to each step and every breath we take. The law states that the total amount of energy in an isolated system is conserved.

Energy is neither created nor destroyed.

Okay what does that mean? Consider a fire. A fire deceivingly makes matter vanish into thin air. The truth is, by the law of conservation, all matter, all energy is conserved. Energy can neither be created nor destroyed; the total amount of energy in the universe is instead finite and conserved. The energy in a log before throwing it to the fire is exactly equal to the energy of its ashes plus the heat and the smoke produced by burning it. The ramifications of such a law are profound. Everything is accounted for and everyone is accountable.

Our perceptions of movement and change are in essence the continuous transfers of energy from one form to the next.

By the law of conservation all motion in the universe is coupled. Sir Isaac Newton recognized this facet of life and eloquently summarized the observation stating in his third law of motion that for every action, there is an equal and opposite reaction. In fact, all three of Newton's laws of motion derive from the single overarching law of conservation. For every step we take, we push down on the ground and the ground equally pushes back up on us!

Similarly, moving mass closer to the center within a rotating system such as the solar nebula results in a counteracting increase in velocity of the system because the total energy must be conserved. With enough increase in speed the outer fringes of the original nebula manages to swirl around the center rather than become a part of it. The increasing rotational speed flattens this remaining mass into a disc of whirling dust and gas.

Newton's Laws of Motion

1. Every object in a state of uniform motion tends to stay in that state of motion unless an external force is applied to it.
2. $F = ma$, Force equals mass times acceleration
3. For every action, there is an equal and opposite reaction.

As for the majority of mass in the solar nebula, it is compacted by the dominating force of gravity into a formidable sphere. The mass that once extended light years across space is condensed uniformly by gravity in all directions into a sphere spanning a mere million miles. Within the massive center, gravity continues pulling mass closer and closer releasing tremendous heat and creating enormous pressure. The pressure increases as more and more mass is compacted into a smaller space. Eventually the forces of pressure pushing outward in all directions exactly equals the gravitational forces pulling inward in all directions and the mass collapse stops. Another balance is finally reached.

The new celestial structure is vastly different from the old. Where once it was light and expansive, it is now dense and comparatively small. Where once it was cold and faint, it is now extremely hot and well defined. In the end despite a polar change in characteristics the structure, like that of its predecessor, remains a balance of two opposing forces, of the inward pull of gravity and the outward pressure of gas. In a universe of conservation and cycles, complementary forces such as gravity and pressure are consistently observed together.

The extreme conditions of the new structure produce a special one-way outlet of energy for the system. Under normal conditions the nuclei of atoms are insulated from each other by the repulsive electrostatic force between like charges. This is exactly the same force observed when pushing the like ends of two magnets together.

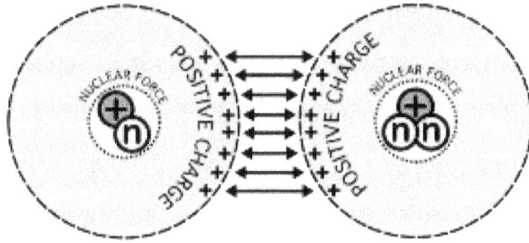

Figure 6: Normal Conditions-- Repulsive Electrostatic Force

The extraordinary pressure in the new structure occasionally overrides the strong repulsive electrostatic force. The force of atomic collisions under high pressure allows the nuclei of two atoms to obtain much closer proximity than normal. At close range, the nuclei of atoms are attracted to each other causing them to fuse into one heavier nucleus thereby creating an entirely different atom! A series of such **nuclear fusion** reactions converts hydrogen atoms into heavier helium atoms. At this point, you may be are wondering how atomic collisions and subsequent nuclear fusion reactions relate to you. Keep reading.

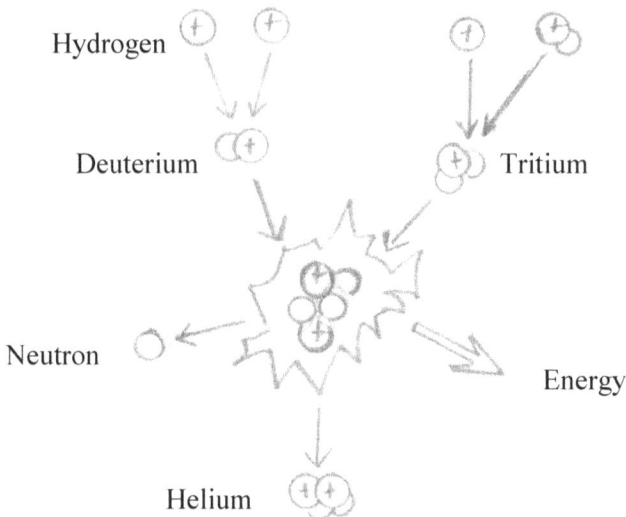

Figure 7: Extreme Conditions -- Nuclear Fusion

Nuclear reactions release a considerable amount of energy compared with other chemical reactions because the binding energy that holds a nucleus together is relatively high. Still, one nuclear reaction, which can release 17 mega electron volts, is hardly remarkable and not even enough to light a light bulb. But, with true strength in numbers billions of nuclear reactions occurring all at the same time radiates an incredible amount of energy--enough to light up the sky. This is exactly what happens in the new structure of the solar nebula and by so doing a star is born—our star.

Our sun is a giant nuclear reactor that converts approximately 700 million metric tons of hydrogen into helium each second. Hydrogen as it turns out is the most abundant element in the universe, making it a good source of long-lasting fuel. The process now nearly five billion years old continues to release nuclear energy to the surrounding solar system, what we feel and perhaps take for granted as the light and warmth of the sun.

One day the sun will inevitably use the last of its hydrogen fuel rendering its structure as we know it unstable. Our star at that point, along with the rest of our solar system, will undergo another transformation governed as always by the law of the universe—the law of conservation. But that day is another story some five billion more years into the future. And to be sure, the day our sun makes the transformation from its star qualities, another star somewhere in our galaxy will be born.

As for that disc of mass that managed to escape the gravitational pull of the sun, it cooled, accreted, and underwent a series of violent collisions and dramatic changes to of course become the eight planets, moons, asteroids, comets and dwarf planets of our solar system. With negligible friction in space and because energy is conserved this mass continues to orbit its star while rotating about itself in near perpetual motion around and around in a 4.8 billion year old whirlpool of outer space!

Like the sun, the planets are in dynamic balance having just enough rotational speed to avoid being pulled into the sun, but not enough speed to escape toward the vast space beyond. The system is both stable and delicate. And as before, the universe repeats itself at different scales; our solar system rotates along with all other mass in the galaxy around the central black hole of the Milky Way.

TWO

Planet Earth

Approximately 99.8% of the mass in our solar system is within the sun. Of the remaining tiny portion of mass, most resides within the two giant gas spheres, Jupiter and Saturn. Put into perspective, Earth is a mere spec of dust and gas. But oh what a wondrous spec of dust it is!

Of the eight planets only one holds the life of the solar system (and of the universe as we know it). Picture the neighboring planets of Venus and Mars with bustling cities like Tokyo and New York. It certainly is a thrilling idea, but no such evolution happened except of course on Earth.

Not only does Earth contain life, it is literally teeming with a wide array and diversity of life. In one small corner of my yard there are microorganisms everywhere sharing space with a range and diverse plethora of complex and organized life. Crawling around in the soil and mulch are little sow bugs and millipedes. Running quickly here, there and everywhere are geckos and cockroaches. Buzzing around in the air are bees and mosquitoes. Reaching to the heavens is a bush of hibiscus, a cherry tomato plant, a cluster of fuchsia bougainvillea, an exotic purple orchid and a towering flower tree with a canopy of leaves. Chirping in the tree are little birds perched on its branches. And of course in my yard from time to time is me, so-called intelligent life!

My yard happens to be on a tiny island in the Pacific Ocean. The range of life in one small corner of my yard is vastly different from another such corner just a few thousand miles from me. Life remarkably crawls, flies, swims, towers, is buried under soil or deep beneath the ocean et cetera in every reach and corner of Earth.

Without a doubt Earth is a special, special place. But what is it exactly about Earth that makes it so bountiful and the rest of the solar system so comparatively barren? A combination of layered conditions orchestrate together to create the Earth's unique music of life. For example, the Earth has a molten iron core and relatively fast rotation about its own axis. Together these two conditions generate a protective magnetosphere not unlike the deflector shields of science fiction. The magnetosphere deflects harmful rays constantly emitted from the sun.

Another condition is Earth's size, which is related to the strength of its gravity. Earth's gravity is strong enough to keep a protective insulation of gases extending some 350 miles beyond its surface otherwise known as our atmosphere. The atmosphere acts as a perfect blanket trapping just enough heat from the sun keeping the Earth's surface at livable temperatures. Of course, climate change may lead to an atmosphere that traps much more heat than we are used to. While both such examples are important to life on Earth, magnetospheres and atmospheres are not unique in our solar system.

Figure 8: Earth's Magnetosphere as depicted by NASA

Undoubtedly the one crucial, masterpiece arising from just the right combination of conditions is liquid water. No other planet has it. Not only does Earth have this unique condition, it has an abundance of it. Owing to just the right distance from the sun, the majority of water on Earth does not boil off into the atmosphere nor does it freeze the world over in extensive ice sheets.

The majority of water on Earth exists as liquid in our oceans. The light silicate crust of Earth bathes in an ocean of liquid water that covers about 70% of its surface. Earth, is a watery "blue planet" indeed. As liquid water is an essential irreplaceable ingredient for life, its special existence alone makes ours possible.

Though abundant on Earth, life is clearly not a given. Life on Earth arose from a solar nebula falling out of balance to create a star, the gravity of which keeps Earth bound and the energy of which sustains life in just the right conditions found only on Earth of the eight planets. Supporting the unique conditions found on Earth, which nurture life is both a means for survival and our birthright. The answers to how we should go about living with Earth are naturally embedded in the operations and cycles which gave rise to us in the first place. So far in our quick study of the birth of our solar system we have already uncovered Nature's three key best management practices for design that supports life.

First and foremost, Nature actively abides by the law of conservation in her design rather than fighting it as our conventional design often does. It may take awhile, but in the end the law of conservation always, always prevails. Therefore, design which works with the operating system of the universe is in our best long-term interest and helps to naturally increase the efficiency of the design.

Under the law of conservation, pairs of opposites are found everywhere. The law of conservation gives way to unity in polarity. Every move, every action, every transfer in energy requires an opposing reaction to conserve the total energy of the system. Because the universe operates under such a rule, Nature uses opposing forces together to do work instead of against each other in conflict. For example, opposing forces are used to create whole balanced entities as large as the solar nebula and the sun to entities as tiny as the atom and the molecule. Opposing forces allow for movement of energy; differences in temperature naturally generate powerful winds.

Second, Nature takes the path of least resistance and minimizes operational costs by maximizing the resources available at hand. The sun, which is the primary source of energy for our solar system uses hydrogen fuel because it is the most abundant element in the universe. Lastly, Nature uses organization and collaboration between individual elements to produce whole entities much greater and more complex than the sum of its parts. For example, the entity of the sun is made possible only through the joint massive effort of its individual atoms. Likewise, our bodies are made possible only though the joint and organized effort of trillions of individual cells. Nature uses organization of the masses to produce impressive collaborative results.

Nature's Three Best Management Practices

1. Design with respect to the law of conservation.

2. Maximize use of readily available resources.

3. Collaborate to achieve together far more than individuals ever could apart.

THREE

A Refreshing Ode to Water

A natural consequence of adhering consistently to Nature's three best management practices is a sense of variations on a theme over and over at vastly different scales. Water, the elixir of life, is one such product of Nature. This is no doubt the reason why water is uniquely essential to life.

Sixty percent made with water humans typically perish after only several days without it. In order to sustain our watery existence, we drink a host of beverage products in addition to eating a host of juicy fruits and vegetables. By extension, water not only nourishes our bodies, it feeds our lifestyles as well. Behind agriculture, industry and energy are the two major users of freshwater.[1]

Fortunately and not coincidentally, the water molecule is the most abundant molecule on Earth. Its abundance owes to its use of readily available resources. The water molecule is a happy marriage between the most abundant element in the universe, hydrogen, and the third most abundant element, oxygen.

As its molecular formula, H_2O, indicates each molecule of water consists of two hydrogen atoms and one oxygen atom. Inherent in the design of each atom, the smallest unit of matter, is the law of conservation. The atom consists of opposing electric charges that we designate as negative and positive. It also consists of opposing characteristics of weight. Protons or particles with positive charge are comparatively massive and contained in a dense nucleus. Protons are approximately 1835 times more massive than electrons! Electrons on the other hand, carry negative charge, are tiny, light and travel at extremely high speeds (only two orders of magnitude slower than the speed of light).

Opposites attract as the saying goes. And as with many clichés, the observation behind them rings true. The electrons of negative charge and light mass are attracted to the protons of positive charge and heavy mass. The attraction binds and stabilizes the two particles into one unit of matter, the atom, the building block of life. Again, in a universe of conservation, pairs of opposites are often observed together as such complementary forces are necessary in any given system even in the smallest unit of matter.

---

[1] 3rd UN World Water Development Report. 2009. Chapter 7 Evolution of Water Use

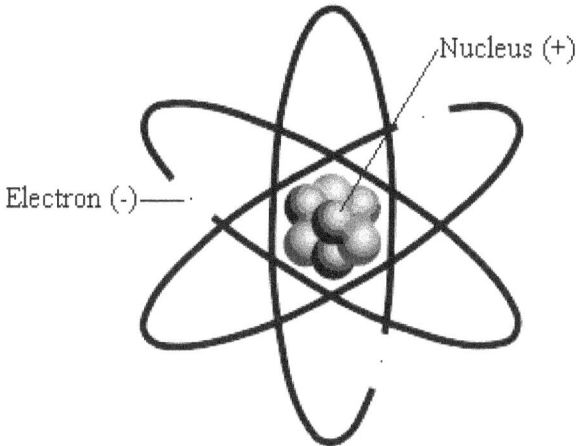

Figure 9: Structure of an Atom

As mentioned earlier, the consistent adherence to Nature's three best management practices results in variations on the same theme. The same design concepts in the atom are used in the construction of all other matter arising from the atom. Particles collaborate to create atoms; atoms collaborate to create molecules; molecules collaborate to create compounds and so on and so forth. We ourselves are the aggregate products of extensive, complex collaboration between trillions of individual cells! The impetus for collaboration is always mutual benefit that cannot be arrived at otherwise.

The water molecule is an example of a collaboration of individual atoms. The one oxygen and two hydrogen atoms in the water molecule physically bind together in order to share electrons, which solves an electron pairing requirement in the world of chemistry. All of the unique and important properties of water derive from its simple yet profound molecular structure.

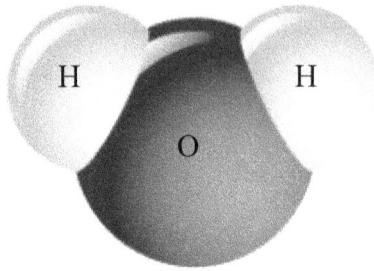

Figure 10: Structure of a Water Molecule

The reason the water molecule is so powerful is unsurprisingly due to opposing features. By weight, the oxygen atom is nearly 16 times heavier than each of the hydrogen atoms. The difference in weight sets up a difference in the way the electrons are shared. The shared electrons are more attracted to oxygen's heavier nucleus making the oxygen atom slightly negative and the hydrogen atoms on the one side of the water molecule slightly positive.

Altogether the molecule is a cohesive and balanced whole while celebrating its bipolar differences! This is the key to water and to the universe—unity in polarity. In each atom and molecule of water, opposing characteristics collaborate and balance to create a functioning whole.

The two distinct negative and positive poles of the water molecule are used for work. Because opposite charges attract and like charges repel all water molecules in contact, no matter how many, systematically bind together and align in a certain way. The attraction between the positive and negative ends of molecules, known as hydrogen bonds, makes water tend to be "sticky". Each water molecule can stick to four other water molecules at a time.

The individual hydrogen bonds are relatively weak and constantly break and reform in liquid water. En masse however, the hydrogen bonds exhibit a much higher collective strength, giving extraordinary unity to whole collections of water molecules. Due to both the individual weakness and collective strength of the hydrogen bonds, liquid water is both fluid and unified at the same time. Mass amounts of water molecules bind together in this fashion to create whole bodies of unified water as small as one glass to as large as the ocean the world over.

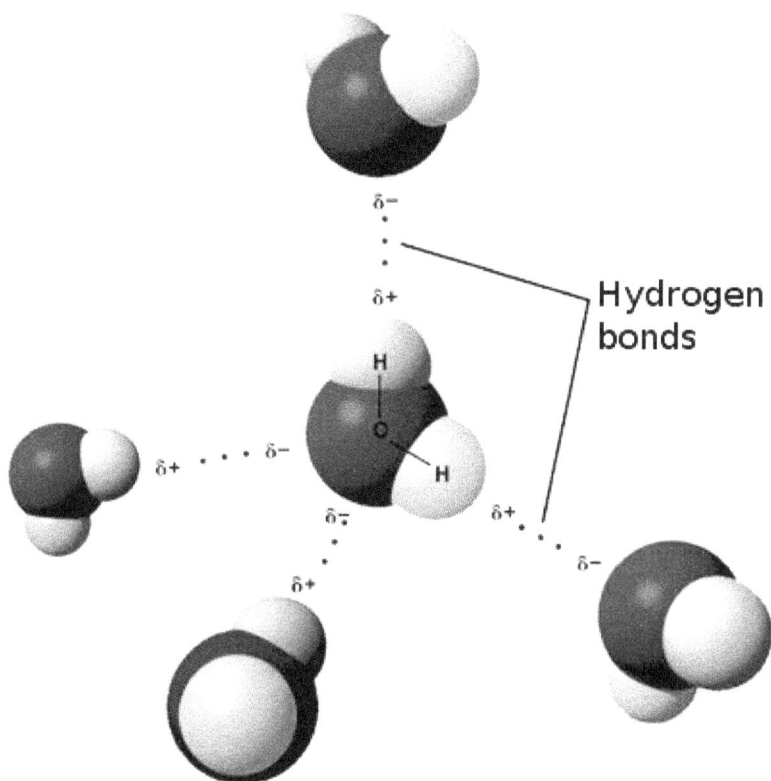

Figure 11: Attraction between Positive & Negative Ends of Water Molecules

The polarity of water molecules is also why water is known as the "universal solvent". Water molecules can surround both positive and negative charged ions, while still linked to other water molecules. In this way, water excels at dissolving and integrating other compounds such as sugar and salt into one aqueous solution. For a compound that is highly unified, it is remarkable that it is also extremely inclusive rather than exclusive.

Building upon the foundation laid by just one water molecule, the water compound is able to harness a collective strength greater than the sum of its individual parts. Comprised of the most readily available elements in the universe and taking advantage of opposing characteristics between individual elements, water effectively demonstrates Nature's three best management practices and by so doing exhibits extraordinary properties of unity and organization.

Water behaves in unexpected ways when compared with other compounds. At cooler temperatures, water behaves like most other liquids—it shrinks. With less energy, molecules move slower and pack closer together resulting in a denser, heavier compound. But pushed to a specific threshold, water does a very curious thing.

Beginning at 4°C and colder to water's freezing temperature at 0°C, water expands rather than contracts as it arranges itself in highly ordered solidarity to an organization that literally floats above expectations. Within these 4 degrees, water molecules transition to its solid form by arranging systematically and freezing into a stable, highly organized three-dimensional pattern of hexagons resembling the structure of diamonds! The hexagon structure of ice takes up more space than the randomly arranged structure of liquid water and therefore is lighter and floats.

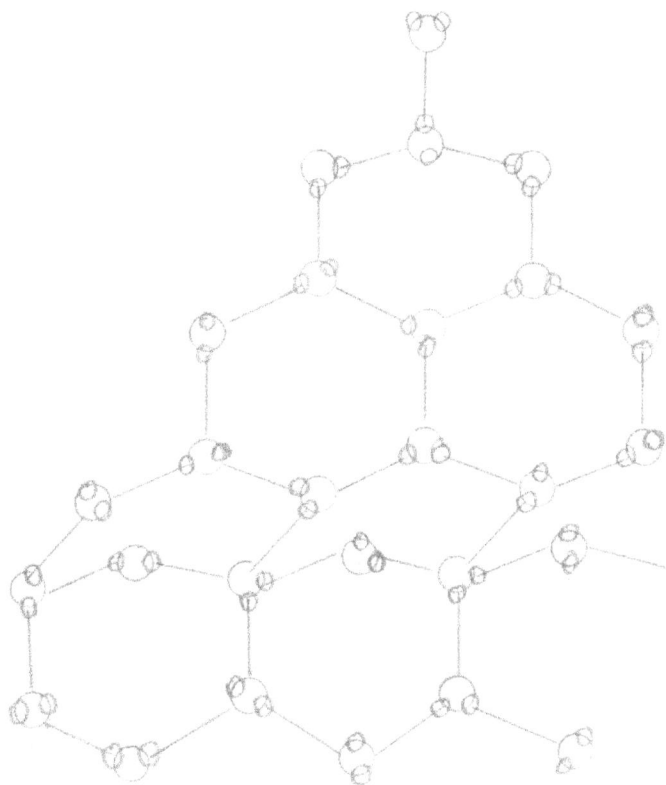

Figure 12: Molecular Structure of Ice

Figure 13: Iceberg in Greenland

The highly organized structure of solid water gives us magnificent icebergs, those massive solid sculptures that despite its irregular shapes and ominous sizes float effortlessly above the ocean's surface. The lack of such organization in other compounds results in a solid form that would sink. The highly ordered behavior of water is also spontaneous. Like jazz, water adheres to certain rules while flowing in spontaneous creativity. The oxymoron of water's spontaneous organization gives us beautiful intricate art in each one of a kind snowflake.

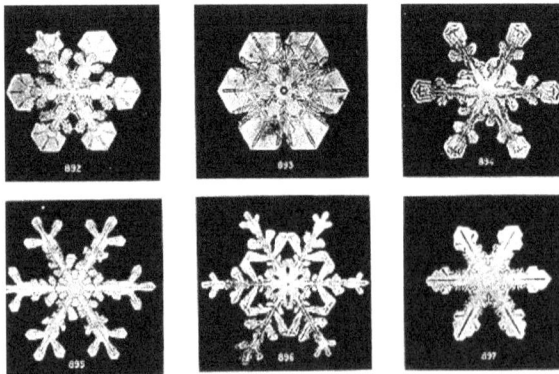

Figure 14: Snowflakes

Water also behaves in unexpected ways at the other end of the temperature spectrum. The unity or strong attraction between water molecules gives water a high specific heat capacity. Compared with all other common compounds, liquid water can absorb the most heat per unit of mass before increasing in temperature. Sand for example increases in temperature five times faster than liquid water given equal masses of each and equal sun exposure.

Water is able to store relatively high amounts of heat before increasing in temperature because of the collective strength of hydrogen bonds between individual water molecules. Before energy is converted to heat, a significant amount of energy is required to first break the hydrogen bonds, which constantly reform in close proximity. Other compounds of similar weight increase in temperature much faster due to less hydrogen bonding between molecules.

The high specific heat of liquid water combined with being the most abundant molecule on Earth makes water an effective coolant for the planet and all of its inhabitants. By extension we use water as an effective coolant for our lifestyles as well, for our cars, power plants, internet servers and for a nice day at the beach. In fact, water's high heat capacity is the reason why water exists on earth as mostly liquid rather than in its other two forms of solid ice or gaseous water vapor. At temperatures naturally found on Earth, liquid water is everywhere— in our deep, vast oceans and cascading waterfalls, in our placid tranquil lakes and gurgling streams, in our cleansing baths and in our life supporting veins.

Given such extraordinary and unique properties, water seems more like a miracle unconfined by the same limits as other compounds. However, water indeed follows the same physical rules as the rest. The difference is that built-in to its design is a full incorporation of Nature's key best management practices. Based on this solid design water exhibits superior unity, organization, collaboration and integration. Out of this shining example of design, life emerged on Earth.

Humanity would do well to mirror the design features of the water molecule and compound in order to produce like-minded results. Temperance under fire would significantly help our efforts toward world peace and unity as one people; as would an unparalleled ability to dissolve and integrate differences into a cohesive whole. Pushed to our own environmental thresholds and limits as humans, we would surely rise above and survive best by organizing together in ice-like mass solidarity.

Part 2

Our New Environmental Paradigm

Three critical management areas requiring an overhaul in best management practices with respect to Nature are: water, energy and byproducts. These areas are globally relevant and are, by our own design, severely out of balance as the writing on the wall reads of water scarcity, energy crisis, climate change and pollution. Crisis as it relates to resources is nature's mechanism for spurring evolution, for extinction of species that are unable to adapt and for growth of species that can embrace Nature's key best management practices.

Our current designs reflect our creative youth and are fairly primitive compared with the complexity of designs demonstrated in Nature. Like weeds, we are obnoxious tending to grow at the expense of other life around us. Eventually because the law of conservation cannot be cheated, what goes around comes around; obnoxious growth has its inevitable consequences. Sure enough, the signs of the time call for our evolution, for a little growing up, for us to live in harmony with our environment rather than in conquest of her.

ONE

Water Management

Water, water everywhere, is inextricably connected to
Water, water everywhere

Nature's Design - the Hydrologic Cycle

As water is essential to life itself, managing our water is of utmost importance. Some liken water to the oil of the 21st century. As the population grows along with its excesses, we fast approach the limits of both fossil fuel reserves and our freshwater capacity. Unlike oil, however, there is no alternative fuel for water and the consequences are much direr. Water is sustenance and irreplaceable.

The presence of liquid water at the Earth's surface is just one part of a larger and much more complex picture. Rain, snow, glaciers, rivers, streams, springs, artesian wells, groundwater, lakes and oceans are all parts of one fascinating dynamic continuum. The entire collection is conserved and naturally recycled over and over. Like all other energy, water continuously converts from one form to another in a larger operating circle. This circle of water on Earth is often referred to as the hydrologic cycle.

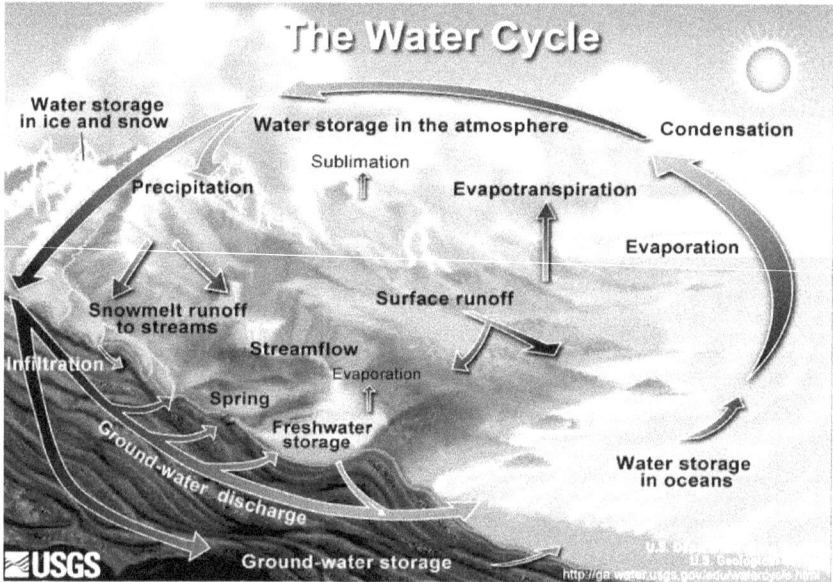

Figure 15: The Water Cycle

The hydrologic cycle incorporates Nature's three best management practices. The cycle is naturally powered by two readily available resources, the sun and Earth's gravity. Not surprisingly because the universe operates under a law of conservation, these forces oppose each other, allowing for movement and balance within one entity.

The sun heats the Earth converting a portion of its liquid water to vapor. With elevated energy, water vapor rises and soars to the sky above. At cooler temperatures high above the Earth the vapor releases its gained heat and condenses back to liquid form. When temperatures are in the freezing range the water droplets turn to ice. Water droplets or ice crystals in the atmosphere tend to collect together forming clouds, which travel with the winds across the globe. Eventually, the clouds collect enough water that gravity naturally pulls the mass back down to Earth's surface as rain or snow.

The amount of time water spends in a particular form varies orders of magnitude. In colder regions on Earth near the polar ice caps, water may remain as ice for tens of thousands of years. That same water may spend thousands of years in an ocean, hundreds of years in shallow groundwater, a couple months in soil, a week in the atmosphere as clouds or a few hours in our body. The same water that ancient Romans drank may very well be the same water we swim in or water our plants with today!

Perhaps because the transport of water from Earth to the atmosphere is invisible, its easy to overlook the connection and believe that clean water will always fall from the sky irrespective of what we do with it on the ground. Not to be fooled, all forms of water are intimately connected. Our use of water affects our reuse of this naturally recycled resource. When we pollute our rivers, without fail we are poisoning our drinking water.

The abundance of water adds to our misperception causing us to take it for granted. In our daily living and by design we largely ignore its value. We squander it, pollute it and even treat it as something to get rid of as quickly as possible--as a flood hazard to public life and property. We use the ocean as a vast receptacle for debris and excess heat. But our water resource is finite and precious. Design which blends our use of the water system with its natural cycle is crucial for our survival. By working with the natural cycle of water we automatically incorporate Nature's three best management practices into our own water management design.

Our Design - Stormwater Runoff

A prime example of where Nature's designs clash with our own is the way Earth processes rain and snow. In the hydrologic cycle, rain and snow are the primary vehicles for returning water in the atmosphere back to the Earth's surface. Nature's design includes a percolation of rain and snowmelt through the Earth's surface, which naturally purifies water to a supply clean enough for us to drink. Our design by contrast includes development, which decreases the percolation of water through the ground resulting in negative impacts to our drinking supply. The misstep in design is further compounded by more misguided design.

Stormwater runoff is a natural process, which in the words of the United States Environmental Protection Agency (EPA) "occurs when…rain or snowmelt flows over the ground." It is the water not readily absorbed into the ground during a storm event—the puddles and little rivers that form only after it rains or snows. Stormwater runoff is considered by the EPA as a "problem", citing that stormwater picks up debris, chemicals, dirt, and other pollutants and transports these pollutants to our lakes, streams and oceans[2]. By the EPA's account and by conventional civil engineering design, stormwater runoff is considered a nuisance, as something which pollutes and is a flooding threat to public property and health. Herein lies a good portion of our distorted thinking. The reason stormwater runoff is a problem is largely

---

[2] *After the Storm.* U.S. Environmental Protection Agency. Web. Sept. 2009

due to our own design. Stormwater runoff picks up and transports debris and harmful chemicals because we generate debris and harmful chemicals and deposit these pollutants everywhere—on our lawns, in our laundry, on our streets, in our businesses, on our agricultural lands, in our factories and so on and so forth.

We "develop" the land with impervious materials such as buildings, pavement and concrete so that a significant amount of rainwater, which used to percolate and filter through the Earth's surface, is no longer able to do so. Then in order to prevent flooding, which our activities tend to enhance, we design an elaborate drainage system of sloped sidewalks, roads, gutters, ditches, channels, drain boxes, and vast piped networks to fast-track the now polluted rainwater resource straight to surrounding lakes, streams and oceans. If that weren't enough we turn around as a firmly established and official institution of a developed country and say that stormwater runoff is the problem! Humans are funny creatures, you have to admit. But it is high-time we put ourselves on the hook and go back to the drawing board.

Not all institutions ignore the elephant in the room. Instead they point right at it then try to carry on in the remaining confined space. The State of Hawaii Department of Transportation (HDOT) openly admits that our designs are problematic. In a manual relating to water quality published in March 2007, HDOT states: "Site developments result in land use and land cover changes thereby altering the hydrologic cycle."[3] The manual goes on to explain how we increase

---

[3] *Oahu Stormwater Management Program Plan*. Best Management Practices Manual. State of Hawaii Department of Transportation. Mar. 2007.

stormwater runoff and pollutants and decrease the natural filtering of rainwater.

Rather than call for human development that does not alter the hydrologic cycle or pollute land, air and sea, the obvious solutions, the manual outlines (over one hundred pages) in depth how to properly treat the symptoms of our flawed design. Large detention ponds designed to capture the increased runoff and settle pollutants are a typical prescribed medication for our drainage ailment. However, these ponds require a lot of precious real estate and need to be dredged to maintain capacity. Moreover the ponds, designed only to control the release of runoff from a site for flooding purposes, do not solve the percolation issue. All of the captured water is instead detained and released off-site over time.

The manual is a prime example of our current acceptance for flawed design and the momentum of an already established system. The manual required and written under the direction of the U.S. EPA is typical of general industry practices in the United States. We can of course continue to treat the symptoms of a flawed design, but in the end the law of conservation always prevails. The call for smarter design which blends seamlessly with the environment is inherently clear. All we need to do is respond with equal volumes of conviction and action.

Where We Need to Be: Rain Gardens

The good news is we have already started the work in small but growing numbers. Progressive cities such as Seattle, Oregon and San Francisco, perhaps pushed along by reaching environmental constraints, are leading the way in implementing more integrated and progressive design concepts in the United States. For example, the City of San Francisco recently released guidelines and design concepts, which strive to 1) blend our activities with, rather than alter, the natural hydrologic cycle and 2) to reduce pollution in stormwater runoff.

Figure 16: Rain Garden in Parking Lot

Design concepts such as *rain gardens*, *green roofs*, and *pervious pavement* seek to restore or maintain the natural filtration of rainwater into the Earth's surface. Relying on natural systems is a way to maximize our use of available resources a key best management practice of Nature. Rain gardens are small garden areas designed to receive stormwater runoff from developed hardscape areas such as roofs and parking lots. The idea is to account for the effects of hardscape on stormwater runoff. Rain gardens are designed to allow rainwater or snow melt that would otherwise flow offsite via streets and gutters to filter through the ground onsite instead.

Rain gardens can also be designed to enhance the natural filtration process and to treat typical pollutants from our activities such as oil and metals. The gardens are designed with native plants, which are the most adapted to the local climate and requires the least maintenance. Working with nature makes life so much easier on all fronts!

Green roofs follow a similar concept on the rooftops of buildings. The idea is to replace the hardscape of our rooftops with plants and a layer of soil to mimic natural landscape, which lives and breathes, which interacts with the surrounding environment. The green roof also provides a layer of insulation for the building, keeping temperatures naturally both cooler during hot days and warmer during winter months. Again working with nature tends to yield a layering of benefits.

Another progressive design solution harkens back to the days of the Roman Empire. Ancient Rome used extensive rainwater harvesting systems in developed areas. Rather than waste rainfall, we can harvest it and use it as a valuable resource. Rainfall can be captured and released gradually onsite as free irrigation water, water for livestock or water for artificially recharging groundwater aquifers.

Since water is essential to human life, paramount to engineering design is design which absolutely supports not disrupts the natural hydrologic cycle. Simply striving to comply with this sustainable principle is not good enough unless of course we are prepared to live with the obvious detrimental consequences. The solution of design that honors rather than alters the time-tested and centuries-old cyclic movement of water on our precious blue planet requires only a modicum of the imagination and intelligence necessary to launch a 4.5 million pound rocket all the way to the moon. Clearly we have the means to evolve, but not the collective will so far. We could wait until the issue of water scarcity forces a change or we could with enough awareness be unable to continue with our current design flaws.

Walking the Talk: Rain Barrels

At my home, rainwater is definitely used as a resource. With a hacksaw to the downspout of my roof gutters, I disconnected my roof drainage from the City's drainage system of roadways, street gutters and concrete channels leading to the ocean not too far away. Sawing metal has never been so satisfying. The downspout now connects to a homemade rain barrel. Rather than buy a commercial made rain barrel, which can cost upwards of $200 plus shipping, I maximized use of local resources. I purchased a used food-grade 50-gallon plastic barrel from the local Pepsi distribution warehouse for $10. Pepsi uses the barrels to store syrup after which the barrels are typically discarded. With a drill

Figure 17: Rain Barrel at Residential Home

and a trip to the hardware store I added a spigot and overflow valve to the barrel and voila! I converted Pepsi's trash to my very own rain capturing device. The rain barrel captures water collected from my roof, which I do not treat with chemicals.

I mainly use the captured rainwater as irrigation for my vegetable and flower garden. The water is free from additives such as chlorine and fluoride and best of all the water is free period. In my city, the wastewater department charges expensive sewer treatment fees for all the water my household consumes, including the water I sprinkle my garden with! Clearly, the wastewater department does not have to treat the water used in my garden. Therefore, using rainwater captured from the roof of my home for irrigation purposes saves me on both my water and wastewater bills.

In addition to a fiscal benefit is an environmental one. By capturing the rainwater from my roof, I am mitigating the stormwater runoff produced from my development of the land. Since I use the water on-site in my garden the water is able to naturally percolate and filter through the Earth's surface eventually recharging underground aquifers.

Where We Shouldn't Venture: Desalination

Some proposed solutions for managing our water resource lack a serious evaluation of our current designs and more importantly disregard the law of conservation. Such solutions include building more damns and desalination, which is the process of separating salt from sea water. As we head into climate change and global warming, the idea of tinkering with the Earth's natural ability to moderate temperatures seems unthinkable. Covering 70% of the Earth's surface, water serves a very useful purpose. Like the coolant in your car that keeps the engine from imploding, the ocean keeps the engine of life operating within its limited temperature range. Solutions to our environmental problems most certainly require cooperation with Nature rather than a continued effort to dominate her.

In trying to solve the impending drinking water scarcity problem, we inevitably look seaward to the obvious conundrum before us. For the sailor lost at sea without a drop of freshwater to drink, the water conundrum is all too stark and desalination is the obvious solution. But the sailor lost at sea is one person of many and the situation is temporary.

Desalination as a permanent and large-scale global solution on the other hand (a solution proposed by many), would certainly throw a wrench into the time-tested cycle of water on Earth. Desalination cannot be a feasible environmental option knowing firmly that by the law of conservation all water on Earth is finite and connected. The current arrangement of mostly ocean water and only about three percent freshwater is a delicate equilibrium found on Earth after four billion years of life supporting capacity. The proposal to alter this traditional arrangement on a large-scale basis is precarious bordering on suicidal.

Despite grave misgivings for the idea of large-scale desalination, the U.S. National Academy of Sciences (Academy), an organization that advises our government leaders on science matters, concluded in April 2008 that "desalination will likely have a niche in meeting the nation's future water needs."[4] This conclusion is reached

---

[4] *Desalination: A National Perspective.* Report Brief. National Academy of Sciences. Web. 16. Aug. 2009.

after reporting "considerable uncertainties about potential environmental impacts", "high costs dealing with concentrated salt wastes", and that desalination is an "energy-intensive process that can add to green house gas emissions." Among the likely environmental impacts are trapping fish and other marine life in intake structures, depletion of groundwater, subsidence (land collapse), negative effects of concentrated salt waste, and diminished water quality and quantity in adjacent lakes, streams and aquifers.

Still, the Academy concludes that while desalination warrants further research, efforts do not need to be halted until such research to better understand the impacts and uncertainties is completed. Yikes! This comes from an organization that advises our government on scientific matters? Clearly, the Academy is more influenced by the momentum of our absurd ways rather than on sound science, making no mention of a comprehensive management scheme, of conservation as a first measure and desalination as a last resort. The Academy is ready to launch full steam ahead without scientific data!

The Academy wishfully states that the, "desalination technology offers the potential to convert the almost inexhaustible supply of seawater…into a new source of freshwater." No amount of research will show that the law of conservation does not prevail. Our planet does have an abundance of seawater, but by no means is the reservoir inexhaustible or "almost inexhaustible". (Is that even a scientific term?) The amount of water on our planet is most definitely finite until a source of water from outer space crashes into Earth and delivers more.

There is most likely a threshold quantity of seawater necessary for the ocean's continued operations as a coolant for the planet and as habitat for a diversity of marine life. Taking a significant amount of water from the oceans does not generate a new source of water and will most certainly adversely affect the surrounding water veins connected to the ocean. Further, having to process our own drinking water supply does not maximize or enhance the naturally filtering resources available to us.

Apparently, the Academy is not conservative with regard to the environment and our most precious water source. We are willing to overlook the negative environmental impacts and high costs of desalination in order to keep the inefficient beast of our ways alive. Where the environment typically helps to shape species evolution, humans insist on standing apart from the rest; we live in a society in which our artificial dominion over the environment prevails. Rather than adapt continually toward best-use water policies, desalination is a means to forge ahead with another feeble attempt to triumph over our physical limits.

The Academy's report is based on findings from the Committee on Advancing Desalination Technology. Where the Academy is concerned advancement of desalination is a forgone conclusion irrespective of scientific findings! Big businesses may attempt to corner the market on such a precious commodity charging high prices for an expensive process that we cannot do without. Before we resort to compounding our water resource problems with more of mankind's altering of the natural cycle, it is well worth examining and improving our current process.

Abiding by the law of conservation--the basic principle that governs life we can and will live in harmony with our planet, particularly with our water resource. The dream is not wishful thinking, but an imperative vision to our survival and represents the next natural progression of our species. There is no time better than now to start giving water our overdue design attention.

TWO

Energy & Climate Change

Our current energy crisis is a calling for our own evolution. Considering the tremendous conclusion from the law of conservation that energy is neither created nor destroyed our energy crisis is not about scarcity at all. Energy is everywhere around us in abundance. The crisis is rather about evolving from using crude methods to the more sophisticated and elegant. It is time to move forward from sucking the ancient sludge of the Earth and burning it to power our modern lifestyles. It is time to incorporate Nature's three best management practices into our energy hungry lifestyles. By will and by necessity we will transition more fully to using readily available energy, to catching sun beams and racing winds and to tapping into ocean waves and waterfalls to power our ways.

Our straightforward and cheap use of fossil fuels was never going to be a permanent solution. Fossil fuels as their name suggests are remnants of a distant past, ancestors that lived and came before us, their energy processed by heat and pressure over hundreds of millions of years into a liquid easily consumed by us for our various energy needs. Nearing the end of supplies for this *easy* energy, we feel a crisis coming on forcing us to grow up and wean ourselves from Mother's milk. We are forced to be more creative and to discover or invent our next fuel cow. By necessity we will get there, but not without crying along the way. Fortunately, life on Earth has a story of evolution spanning billions of years from which we might learn a thing or two about converting energy for our own uses and save ourselves a few tears.

One of the original inhabitants of Earth and from which all other life evolved is a group of simple yet profound, single-celled organisms called the prokaryotes, which includes bacteria. Prokaryotes inhabit the Earth everywhere, making use of a diversity of readily available energy sources. They are found in the air, on the surface of the ocean, deep beneath the sea, in the soil, in our eyebrows, on our countertops and our remote controls.

Equally remarkable to their diverse use of natural resources is the prokaryotes successful adaptation to climate change. As suffocating as it may be to imagine, Earth's early atmosphere contained rare amounts of oxygen. The prokaryotes emerged under these anoxic conditions. Naturally, since oxygen was not readily available the first prokaryotes were anaerobic or did not require oxygen to survive and in fact oxygen was poisonous to them!

Obviously Earth's atmosphere, and life along with it, changed dramatically at some point. Rather than from cosmic forces beyond, the dramatic change is attributed to life itself, to the activities of a particular organism here on Earth. The award for generating the single most profound change for life on Earth goes to the world's smallest organisms, to the ancient bacteria known as cyanobacteria. The widespread growth of cyanobacteria is credited with giving rise to the oxygenated atmosphere we are accustomed to and rely on everyday, every breath of our lives.

Cyanobacteria absorb energy from the sun and use it to convert water (H_2O) and carbon dioxide (CO_2), which is known as a green house gas, to carbohydrates (-CH) and oxygen (O_2). In a nutshell, cyanobacteria are solar powered, drink water, breathe CO_2, produce sugar and fart oxygen. Beautiful creature! It's no wonder why cyanobacteria flourished at conception billions of years ago and are still widely present today. Theirs is a classic energy model that remains very relevant even in modern times. Their life's work continues to be the base of the major food chain, the foundation upon which most other life depends including humans. Considering the tiny one-celled simplicity of cyanobacteria, the ramifications of their activity are incredible. En masse, with true strength in numbers cyanobacteria achieved far-reaching, long-lasting, life-changing effects on Earth.

Cyanobacteria persist in the environment today, often referred to as blue-green algae. They are also present in the chloroplasts of all green plants, serving as energy capturing units for more complex organisms in exchange for a home to live. Scientists refer to this kind of mutual benefiting relationship between organisms as endosymbiosis.

Before life adapted to an oxygenated environment, the rise of oxygen in the atmosphere posed a serious problem to the first prokaryotes. Oxygen as it turns out is highly reactive, causing rust in the presence of iron and water. Similar oxidative chemical reactions cause damage to living organisms. The paradox is that the highly reactive nature of oxygen can also be put to good use. Using oxygen from an energetic standpoint in chemical reactions is much more efficient than not. Prokaryotes evolved to do just that, incorporating oxygen into their metabolism and making use of readily available resources. The higher efficiency has resulted in a predominance of organisms with aerobic versus anaerobic metabolism, including you and me. We are by comparison to our single-celled prokaryote ancestor very complex assemblies of trillions of cells. But perhaps we would not have evolved to such complexity due to energy constraints; perhaps we wouldn't exist at all if our ancestors had not made the leap to the more efficient metabolic system.

As for oxygen's toxic effects prokaryotes adapted; they developed special enzymes, called antioxidants to stop oxidation reactions from damaging cells. The protection allows us to breathe and use oxygen for protracted periods, for whole lifetimes. But the protection is not full-proof; signs and symptoms of aging are attributed to oxygen's damaging effects despite the work of antioxidants. Perhaps at times you wonder why life feels like a paradox; rest-assured we are the living, breathing products of paradox. We cannot survive for very long without breathing oxygen in, but at the same time we are mindful to drink our cupfuls of tea, eat heaps of berries and chunks of dark chocolate all touted to be rich in antioxidants to help keep oxygen's aging effects at bay.

Today's energy transition from fossil fuels to renewable fuels represents a similar increase in energy efficiency as the transition from anaerobic to aerobic metabolism. Relying on the generation of fossil fuels, a process that requires hundreds of millions of years is simply not going to work for our already immense and burgeoning energy needs.

Just as the prokaryotes adapted to incorporate resources at hand that boosted operational efficiency, we too can adapt and make the necessary infrastructure changes to harness readily available *clean* energy.

Where We Shouldn't Venture: Nuclear Energy

Figure 18: Nuclear Power Plant in Cattenom, France

Some, including United States President Obama, support using nuclear energy as an alternative major source of power. Support includes the recent $8.3 billion in federal loan guarantees for the construction of two new nuclear power plants in Georgia. The two new plants would add to the 102 nuclear power plants currently operating in the United States.

Nuclear energy is touted as so-called green energy because it does not release carbon into the atmosphere. However, carbon itself is not the problem. It is the excessive concentration of CO_2 in the atmosphere that is the problem. The byproduct of nuclear energy on the other hand is high-level radioactive waste that cannot be recycled and requires a permanent disposal site. By comparison, carbon is ultra green especially if we incorporate end uses for carbon to keep it from accumulating in the atmosphere. Furthermore, nuclear power plants pose the kind of meltdown risks, already realized at Chernobyl and Three Mile Island.

Added to radioactive issues is the fact that nuclear energy is not renewable. The main fuel used to generate nuclear energy is uranium. Similar to fossil fuels, the supply of uranium does not renew itself at the rate we want to use it. Based on current supplies and demands of uranium, the supply may last the next one hundred years. New technology may be able to stretch that supply to thousands of years. But long before then, we will run out of places to safely store the toxic radioactive wastes on our home planet, Earth.

The bottom line is nuclear power plants do not incorporate Nature's key best management practices and represent an effort that duplicates the sun's generation of nuclear energy dangerously close to home. Further, the Earth's magnetic field and atmosphere naturally help to screen us from the radioactive effects of the sun's nuclear energy. Maximizing use of our natural resources is our best ally in our new environmental paradigm. Our efforts are better spent on designing ways to harvest, store and distribute the energy at hand rather than reinventing the wheel or in this case the sun with potentially disastrous consequences.

Where We Need to Be: Renewable Energy

**Figure 19: Stirling Dish System in Operation at the
Platforma Solar de Almeria, Spain**

Our efforts are best spent on solving the biggest problem currently facing the renewable energy industry, which is converting the wild variability of solar and wind power into a stable, controlled source. Cloudy days do not generate much solar power and at night solar power is of course completely unavailable. The solution is to store the power in batteries for controlled use on demand. This is easier said than done at the moment. If you think about the batteries in your remote control or even the battery just to start your car, you can get an idea of the giant-sized batteries you would need to power an entire household. Batteries which store electricity represent expensive on-going costs. The solution calls for more clever ways to store solar energy and to redefine our notion of batteries.

Again, looking to Nature gives us a cheat sheet if you will for solving our energy problem. The predominant energy continuously supplied to Earth derives from the sun. So it isn't surprising that at the base of our food chain is the solar powered cyanobacteria model—simple and abundant organisms who catch sunlight. Nature's solution for storing solar energy is a two step process of batteries, different from the conventional concept of batteries.

Photosynthetic organisms capture solar energy when it is available and store it in carbohydrates or plant food (first step). A variety of crops all over the world ensures a stable source of plant food. Heterotrophs, like us, eat food ultimately derived from photosynthetic organisms in order to obtain energy. We do this on demand. Our bodies convert the solar energy, which is stored in foods to solar energy stored in the molecule called adenosine triphosphate or ATP, the enzyme that acts as the batteries of our cells (second step).

ATP is an efficient unit of energy for our bodies since the amount of energy stored in one molecule of ATP closely matches the energy needed for most cellular functions. ATP is continually recycled like rechargeable batteries. The power plants which synthesize ATP are decentralized meaning each cell has its own energy factories, called mitochondria, as opposed to huge central power plants which deliver energy across great distances. Incidentally, mitochondria synthesize ATP by using the law of conservation to its advantage, employing the use of opposing forces.

Following Nature's lead, photovoltaic cells, which mimic photosynthetic organisms, should convert captured solar energy to "food" first rather than directly to electricity. The "food" can then be used later on demand to generate electricity. Currently, we use fossil fuels (ultimately derived from solar energy) as the food for electricity. We burn fossil fuels on demand to generate heat, which in turn is used to produce steam. The steam does work by turning turbines, which is used to generate electricity.

One of the best proposed solutions to date is to store solar energy in the form of heat. Similar to food, heat can be used on demand to generate a stable source of electricity via the well devised system of steam and turbines. Rather than having to burn our food, solar power can be easily stored as heat eliminating the negative effects associated with burning fossil fuels.

Clearly, the renewable energy storage solution needs our collective intelligent attention now. And producing this solution is a much better use of our efforts than solving the problem of storing radioactive wastes as required in the nuclear energy scenario. If we can figure out how to split atoms, we surely can solve our energy battery problem.

One possible solution may be to enhance Nature's designs rather than to reinvent it. Instead of using fossil fuels, we can grow and harvest biofuels (fuels generated by living organisms as opposed to dead ones) leaving fuel production with the organisms that do it best. The plants with the highest potential for such an endeavor are none other than the modern day descendants of the cyanobacteria—the blue-green algae. The giant oil company, Exxon Mobile also sees the potential having invested $600 million in 2009 in the world's smallest plants for the production of fuel.

Since the objective is to harness solar energy, algae have an edge over other more complex plants such as corn or trees. Cyanobacteria remain as the world's simplest, smallest photosynthesizers. As such, given the right conditions algae are extremely efficient and can grow rapidly, doubling its weight several times a day. The right conditions for algae involve having an abundant supply of CO_2 and nutrients. As it turns out human activities result in both an abundance of CO_2 and nutrients. Concentrated CO_2 from the stacks of power plants can be used to increase the efficiency of algal growth and to help maintain balance in the carbon cycle. Of course, CO_2 is released when the biofuels are burned. So algae biofuel farms do not represent a net sink for CO_2, but they do provide a continuous way to utilize CO_2 as fast as it is released ensuring no further contribution to the accumulation of CO_2 in the atmosphere.

Likewise, the source for nutrients can be supplied by another untapped and abundant byproduct of human activities--treated wastewater. Algae, unlike other plant crops, can survive in and utilize treated wastewater to grow fuel. Stripping wastewater of its abundant supply of nutrients provides further treatment for our wastewater as it re-enters the natural water cycle.

Marrying the need for energy with the natural sources for that energy is exactly the kind of cyclic design necessary for our new environmental paradigm. Current design practices might use freshwater instead of wastewater to grow algae. Nutrients would have to be added to the freshwater in the form of a costly manufactured commercial fertilizer. Instead, using cyclic designs as Nature does maximizes the use of natural resources at hand, minimizes our production effort and helps the flow of energy. Many of our environmental problems are due to an accumulation or imbalance of energy i.e. pollution, climate change, waste.

Production costs for algae biofuels may be even further reduced if the industry can figure out how to extract the biofuel without killing the algae. The relationship would mirror the endosymbiotic relationship that terrestrial plants developed with cyanobacteria. The cyanobacteria or algae produce food for its host in exchange for a place to live, completely stocked with CO_2 and nutrients. Nature relies on endosymbiosis or collaboration of individuals for efficiency and so should our design for a new environmental paradigm. Maintaining algae rather than having to continuously grow and decimate algal populations would help us cut down production costs significantly.

The invention of algaculture for fuel could very well be the modern complement to our invention of agriculture for food. Our reinvention of the way we harvest energy from the environment most surely includes a revamping of an old invention. Instead of cultivating and farming food crops we will cultivate and farm oil crops. Rather than hunting for oil buried deep within Earth, we will domesticate our oil crops for easy and reliable access. The crops will be diverse; one will be wind; another solar and yet another biomass. Fuel crops will range in size from small residential gardens to industrial size farms. We have the ingenuity and imagination to make renewable energy work. What it comes down to, is our collective will and whether we are committed to the task at hand well before crisis or whether we prefer to wait and suffer certain consequences of the last minute agenda.

Walking the Talk: Photovoltaics

Recently, my household heralded 2010 with a huge energy investment. Rather than just writing about the importance of our energy transition I should be (in theory) one of the first few willing to put money down toward action. I love the idea of generating enough electricity to power all of my home energy needs with readily available renewable energy located on-site. The area I live in, Hawaii, happens to be ideal for capturing and using solar power.

Firmly committed to making the transition to renewable energy, my household purchased and installed photovoltaic cells on the roof of my house. The system currently generates enough electricity to power all of our home energy needs for lighting, hot water, electronic gadgets, cooking on an electric stove and oven, drying clothes, refrigerating food etc. Perhaps in the near future with the purchase of an electric vehicle we will be able to power our transportation too.

Sounds nice, right? Here's the catch. As mentioned earlier, the renewable energy sector needs a battery design solution. Currently, the solar storage options for the average homeowner including my household are to invest in giant expensive batteries or remain connected to the electric utility company. My household opted to remain connected to "the grid" until better options are available.

For now, we transmit our variable solar energy as it is generated during the day back to the electric utility company. The utility company is better able to use the variable energy because of its vast customer base and because most of its energy generation (via burning fossil fuels) is controlled. However, the variability does pose a problem for the utility company who can only feasibly accept 1% variable renewable energy out of its total energy output.

In return for the solar energy, the utility company offers me a one for one trade of reliable electricity available on demand. The service is not free and costs about $20 a month, but it is a seemingly better option than investing in giant batteries especially if better storage technology develops rapidly soon, which is of course what I am hoping for.

At the end of the day, my household's energy transition is a hybrid one much like hybrid electric cars. As technology is developed, helped along by our collective will, we can proceed with full transition and complete the weaning process. The capital cost for the hybrid transition? The out of pocket expense for the project before tax incentives and rebates is an eye-whopping $26,000! The tax incentives are substantial and represent the United States' commitment to energy transition. Without incentives we would not have contemplated such a capital endeavor. Still, the cost after tax incentives of about $9000 is a bit daunting for our middle class working household, which includes children. Having full faith in the need to switch to renewable fuels and having the cash on hand allowed us to push forward past the sticker shock. The investment, in theory, should pay for itself in the long run, but lets be honest the capital cost is a hurdle even for those who can afford it. By necessity we will overcome the hurdle eventually, but if we can summon our collective will and unite across the masses we can with true strength in numbers overcome the hurdle much more easily and quickly together than not.

Climate Change

As for today's changing climate our only solution is to adapt. The accumulation of green house gases, namely CO_2, is attributed mainly to our own activity. We release tremendous amounts of CO_2 in the atmosphere without any activity that opposes or balances that release. As we have already discovered, balanced dynamic systems require a pair of opposing forces. We actually exacerbate the issue by decreasing plant life on Earth in favor of our hardscape development.

Inevitably CO_2 accumulates in the atmosphere as a result, which as we are finding has a negative greenhouse side effect.

Wait. A pattern is emerging. The compounding effects of human activity with regards to CO_2 accumulation sound awfully familiar to the issues faced in another important cycle—the water cycle. Our designs are narrow with little accounting of its effects on the larger scheme or incorporation within the larger cycle. And yet we clearly operate within a universe governed by cycles, by the law of conservation. It is time for us to accept the way the universe operates and to use this knowledge to our advantage rather than to insist on being apart from it.

CO_2 in fact is one of the main ingredients for the photosynthesizers of the world. Representing the base of our food chain, it is imperative that we support the activities of the photosynthesizers. Just as prokaryotes recognized oxygen as a major energy source and incorporated it into their metabolism despite oxygen's very toxic effects, we can recognize CO_2 not just as a greenhouse gas, but as a valuable resource.

Many of the carbon capture proposals identify CO_2 as a waste product that needs to be stored similar to many of the waste products we store in landfills. Waste however, as we define it only accumulates and is unproductive. Our end uses for CO_2 must be incorporated into another cycle. Perhaps one of the best immediate solutions to capturing carbon from the atmosphere is to launch a massive effort to restore much of the biomass that we have cut down or destroyed. Biomass production and subsequent incorporation into soil and sediments is a key pathway of the current carbon cycle.

THREE

Byproducts

Currently, the by-products we generate from our daily activities persist in their original forms long after we release them to the environment. Styrofoam take-out containers, grocery bags, plastic forks, magazines, napkins and coffee cups are among the endless array of items we use and dispose of on a daily basis. We dispose this vast array of items in overwhelming volumes of proportion. This is the human invention called waste, unique among all other byproducts of life. But in a universe of conservation energy is not wasted, it is constantly reused.

By design our waste accumulates rather than being readily reincorporated into the environment causing a significant waste problem for all life. Considering the fleeting and transient nature of our disposables, we are over-producing and over-designing by a long-shot leading to extraneous waste management effort. Our energies are definitely much better spent elsewhere.

Contrast our daily byproducts with the opposite extreme of the spectrum. The oxygen that cyanobacteria throw away happens to be the vital breathing element on which our lives along with most others depend. Imagine our byproducts being as vitally useful to others. Actually, the CO_2 we respire is exactly the gas that cyanobacteria and plants require in photosynthesis. This exchange is just one example of the eloquent, balanced, give-and-take relationship so often found in Nature.

Imagine the possibilities of human design which follows Nature's model; imagine paper coffee cups doing double-duty, serving as both beverage containers and as readily digestible food for other organisms once discarded. The vision is a little blurred; our society having invented and accepted the concept of waste, requires a trip down the rabbit hole to envision a society that does not produce the day-to-day waste we are accustomed to. Still the vision is there. And everywhere around us models of this concept abound.

By necessity, steps have been taken to diminish our proportions of waste. We have launched massive and expensive recycling programs, which make us feel better about our over-designed packaging and give us a good excuse to keep it. The truth is much of our daily disposables remain un-recyclable. And of the waste that can be recycled, these need to be collected, sorted then processed for reuse. As well-intentioned as our recycling programs are, our waste problem calls for a much higher commitment.

The solution is a global change in thinking; we invented waste and by necessity we can relinquish its concept altogether. Where packaging is unnecessary, we should remove it. Of the packaging that is necessary, it should be designed to be readily reincorporated in the environment by natural processes with the least amount of intervention.

The goal may seem impossible. And yet, of all the processes and life forms on earth, humans alone carry the unworthy distinction of creating the idea of waste—byproducts that are not readily consumed by other organisms and therefore, persist in the environment in its original form far beyond their useful lives. All other life on Earth produces volumes of biomass without producing an ounce of waste! Nature gives us a plethora of design models to follow, packaging its short-term products, literally the fruits of its labor in appropriate organic containers of all shapes, sizes and colors. Bell peppers and cucumbers have a waxy, shiny, plastic-like coating that is durable, completely biodegradable, and even edible! Palm trees package their highly prized coconut meat and milk in a sturdy, husk product. Peas are delicately packaged in air-filled, cushioned pods. The embryos of chickens are delivered in delicate yet hard-cased shells. Nature appropriately designs packaging for the purpose and use of a diversity of products. This without fail is the model we must follow for all of our products—appropriate design incorporating Nature's key best management practices.

Where We Need to Be: Compostable Products

A number of our design processes are changing to reflect the ephemeral nature of disposables. We are realizing that by design, waste has no where to go making it one of the larger banes of society.

Following the idea of food for other organisms, designers have developed the idea of compostable products. Compostable products are short-term items designed to readily disintegrate and biodegrade within two months. Derived from plant materials such as corn, these products can be safely consumed by microorganisms. The key to these products of course is to follow through with successful composting, hence the name. A host of compostable products are replacing the era of disposables. Durable yet biodegradable plates, bowls, take-out boxes, utensils, cups, garbage bags and packaging materials are available on the market. The diversity of plant materials lends to a matching diversity of colors and textures in compostable products including a perfect resemblance to clear plastic cups.

Compostable products are verified by the Biodegradable Products Institute (BPI), an independent, non-profit association of individuals, government and industry in the U.S. BPI certifies that products can be successfully composted in a professionally managed composting facility. Of course, successful composting does not require a professional facility. But successful composting, like baking a cake, requires some effort and a few key ingredients such as aeration and moisture. BPI lists their approved products on their website and certified products are labeled with their compostable label.

The whole heap of daily packaging can be efficiently processed not by human hands, but by a readily available labor force--microbial labor. Microbial labor is our most under-rated and untapped resource in recycling products. Microbes work for free, onsite, and round the clock without complaint. Imagine if we designed our waste to be readily reduced by this mass labor force. Contrast this service with the tax-paid garbage collectors of every municipality that haul garbage from every nook and corner to one of few landfill locations. Any way you slice it, microbes are efficient and low maintenance requiring only rainwater, air and organic byproducts. With design we can partner with microbes to effectively recycle the packaging of our products in a timeframe that better matches the life span of the packaging.

With design, our packaging just as the packaging in nature would be completely biodegradable, discarded after use and become food for other organisms within say two months. Of course our new designs are not an excuse to live a "throw away" lifestyle. Where practicable reusable dishes are by far the preferred choice over disposable dishes.

The recycled *end-product* from our microbial partners is very useable organic soil for our flower and vegetable gardens. Rather than expend the energy to drive a car to the local hardware store to pickup a plastic packaged bag of soil that was transported who knows how many miles, you could literally walk out your back door and scoop locally grown, package-free organic soil. Why work harder than we need to?

Walking the Talk: Vermicomposting

This past Easter I celebrated spring with a sprucing up of my garden and a purchase of a quarter pound of worms plus associated microbes. Yes you read that right, "a quarter pound of worms." I am rather squeamish of legless creatures particularly snakes and worms.

However, convinced (academically anyway) of the need for others to make use of my end products I purchased creatures of the underworld specifically evolved to do just that--earthworms. Known as vermicomposting, certain types of worms are used for the efficient processing of compostable products such as vegetable and fruit scraps. The end product of that process is worm poop or vermicompost if you prefer, which is an excellent nutrient-rich fertilizer and soil conditioner that I can use in the garden.

I keep the worms in a bin like pets. I use shredded newspaper as bedding (a free publication is mailed to my home every week). I feed the worms with an assortment of vegetable scraps, fruit peels, old bread and uneaten rice that would otherwise be thrown in the trash can. I would love to add an assortment of compostable packaging products like wrappers and coffee cups to the worm bin as well.

Because the worms require a moist environment, I also water the bin. The water I use is yes, rainwater collected from the roof of my home. At this point I am euphoric with a sense of cyclic being and I haven't even harvested my first yield of vermicompost yet. Kidding aside, incorporating cycles into our daily living is definitely where our new environmental paradigm is at.

FOUR

Collective Brain

The whole is more than the sum of its parts.

--Aristotle

Nature's third key best management practice is collaboration between individual elements to create a whole greater than the sum of its parts. Vital to our successful implementation of our new environmental paradigm is a collaboration of all human efforts to transcend the individual entities. Bound by the human spirit and condition, which absolves cultural and physical borders, we collectively call planet Earth our home. Technological advances in communication help us to create and sustain our next greatest invention—the collective human brain.

For a long time Earth was dominated by the single-celled organism. Only relatively recently did Earth become inhabited by the diversity of multi-cellular organisms that we are familiar with today. Collaborating together en masse is not inherently apparent. The sense of the individual, of the ego is strong. But once single-celled organisms finally learned to organize and work together to create whole entities that could not be created otherwise, a Cambrian explosion of endless possibilities and combinations of cells followed. Similarly, like the trillions of individual cells that transcend their own existence by collaborating to create the human entity, we as individuals can collaborate to seven billion plus strong to create an entity which transcends the human condition. As this one entity we can collectively achieve together much more than we ever could apart and finally move the needle of environmental progress closer to where it needs to be.

Wikipedia, a web-based free content and openly editable encyclopedia where contributors write collaboratively and without pay is a shining example of just one possibility of collective organization and the direction we need to go. Its creed of an organized humanity collaborating together for the greater good where no one person profits above the whole is exactly the business model that made the Cambrian explosion a success. Nearly all of the images provided in this book are by contributors who freely published their works on Wikipedia. If it were not for the collaborative spirit of the Wikipedia model, this publication would not contain images period much less created from authors around the world. Thank you Wikipedia! We have truly arrived at a place for an amazing sharing of information.

According to Wikipedia, it has grown rapidly since its inception in 2001 into one of the largest reference websites in the world. As of January 2010, the website attracts nearly 68 million visitors a month. More than 91,000 contributors are working on over 15 million articles in 270 languages. That people can organize together in such numbers all over the world to create a widely visited and successful encyclopedia for the greater good is a testament to the emergence and power of our collective brain.

The Eco-Patent Commons launched in 2008 by IBM, Nokia, Sony, the World Business Council and others is a another example of collaborative intelligence for the common good of all. The Eco-Patent Commons is specifically geared toward environmentally friendly technology. It is a vehicle for "IP-free" technology or intellectual property feely available to public domain as opposed to patented technology owned by individual businesses. With this approach, development of technology becomes a collaborative effort across industries leading to enhanced, innovative products that would not otherwise be possible. Sharing technology helps to unify our efforts and to make us collectively stronger and more effective. This is exactly the kind of mechanism we need for approaching our environmental challenges as one world-wide intelligent community.

According to the World Business Council for Sustainable Development, a global association that administers the Eco-Patent Commons, one hundred eco-friendly patents by eleven companies have been pledged since its conception.[5] Patents include a wastewater treatment method developed by Fuji Xerox and a more efficient process for producing olefins, a common component in packaging by Dow. Although one hundred patents and eleven companies are a start and an achievement in itself, the numbers are hardly effective or impressive especially compared with Wikipedia numbers. The potential for so much more is evident. More efficient ways to harvest and store solar

---

[5] *Eco-Patent Commons.* World Business Council for Sustainable Development. Web. 15. Apr. 2010.

energy and ways to create compostable products are among the chief technologies we need to focus on collaboratively.

Like Wikipedia articles that focus on a specific topic and are accessible to the world for collaborative creation and use, we need similar spaces for environmental solutions. One *article* would be entitled 'Stormwater Runoff', another 'Biofuels' and another 'Carbon Dioxide Reuse'. Talent from all over the world could focus and collaborate on the greatest environmental opportunities facing us today. Together we can and will accomplish more than we ever could apart. Let's start capitalizing on the huge potential for addressing our collective challenges with our collective brain today.

CONCLUSION

L ife as we know it is unique to our home planet, Earth. A special combination of conditions gave rise to the original inhabitants and these special conditions persist today allowing for modern life to continue nearly five billion years later. Outside of these special conditions, life itself has dramatically altered the landscape and surrounding biosphere. Starting with tiny single-celled organisms that oxygenated our atmosphere through to today where human activities concentrate carbon dioxide in the air, life has the ability to change the conditions found on Earth. Though we have demonstrated amazing feats of design and conquering of our environment, paramount to our survival is a responsibility for how we affect our surroundings.

Today's climate literally calls for a global change in the way we do business with respect to environment. The good news is we are responding to the call, applying our ingenuity in order to keep our modern lifestyles while living *with* Earth. Going green these days is a common practice and even fashionable. But without unifying guidelines these efforts, as well-intentioned as they are, lack effect.

With nearly five billion years of design experience, Nature gives us three unifying best management practices as a guide. First and foremost, the universe operates along the law of conservation. What helps the world go round is definitely design of systems that do the same thing. Living with Earth is accepting the cyclic nature of our universe. Within this operating system, pairs of opposites like the binary system of zeros and ones are used to balance all actions. Secondly, Nature uses readily available resources. Solar power is chief among the most readily available resources used by nearly all life. And lastly, Nature uses collaboration between individuals to create entities much greater than the sum of its individual parts. From the microscopic building block of the atom to vast collections of atoms and cells within the human entity, Nature uses collaboration everywhere.

Along the guidelines of Nature's three best management practices, concepts that do not embrace the law of conservation or make use of readily available resources such as nuclear power plants and desalination should be shelved. Whereas the concept of algae biofuel crops fed with CO_2 from power plants and nutrients from wastewater need to be implemented in full force today. For greater impact and effect, information sharing and collaboration of ideas must be incorporated in generating solutions for the environment. The idea of intellectual property which benefits and belongs to the public domain is exactly the kind of mechanism that will get us there.

Our evolution (or extinction) has always and continues to be excitingly a joint venture between Nature and our own free will. Famous or infamous for our ingenuity and inventions, it seems our free will comes with responsibility. The evolution of man may indeed have a larger purpose. With true strength in numbers our response to the environmental cues observed today includes our next greatest invention—the collective mind. Our various hi-speed inventions for communication coupled with the challenges we face as a people points to a collaboration of minds that transcends the individual. Using similar creative and collaborative spaces as Wikipedia does, we can amass the enhanced strength and ingenuity of a unified humanity required to adequately address our global environmental issues. Like the Cambrian explosion that launched organisms from simple to diverse and complex, we are on the verge of an explosion of collaborative designs that marry human activities with Nature. Unified and organized by the collective mind we will achieve together much more than we ever could apart.

Figure Credits

Listed in order of appearance

1. Jones, Naia. Far Beyond Nomadic Living. 14. April. 2010.

2. S 400 Hybrid (Wikipedia Author). Toyota Prius III. Digital Image. *Toyota Prius*. Wikipedia. 10. Jul. 2009

3. Deretsky, Zina. Fossil fish bridges evolutionary gap between animals of land and sea. Digital Image. *Tiktaalik*. Wikipedia. Public Domain.

4. Hester, J. and Loll, A. A Giant Hubble Mosaic of the Crab Nebula. Digital Image. *Hubble Site*. NASA. Public Domain.

5. Kuntz, K., Bresolin, F., Trauger, J., Mould, J., Chu, Y.H., Hubble's Largest Galaxy Portrait Offers a New High-Definition View. *Hubble Site*. NASA. Public Domain.

6. Panoptik, I. Diagram illustrating in a schematic way the technical difficulties of nuclear fusion. Digital Image. *Nuclear Fusion*. Wikipedia. 3. Aug. 2007. Creative Commons Attribution ShareAlike 3.0. See terms under Creative Commons Section.

7. Jones, Naia. Nuclear Fusion. 14. April. 2010.

8. NASA. Digital Image. *Image Gallery*. NASA Heliophysics.

9. Halfdan (Wikipedia Author). Stylised Lithium Atom. Digital Image. *Wikipedia*. Wikipedia. 18. Mar. 2007. Creative Commons Attribution ShareAlike 3.0. See terms under Creative Commons Section.

10. Sakurambo (Wikipedia Author). Structure of the water molecule. Digital Image. *Wikipedia*. Wikipedia. 10. Dec. 2007. Public Domain.

11. Qwerter (Wikipedia Author), translated by Maňas, Michal. Model of hydrogen bonds. Digital Image. *Hydrogen Bond.* Wikipedia. 3. Dec. 2007. Creative Commons Attribution ShareAlike 3.0. See terms under Creative Commons Section.

12. Jones, Naia. Molecular Structure of Ice. 14. April. 2010.

13. Hansen, Kim. Iceberg near the Harbor enrance in Upernavik, Greenland. Digital Image. *Water.* Wikipedia. 11. Aug. 2007. Creative Commons Attribution ShareAlike 3.0. See terms under Creative Commons Section.

14. Bentley, Wilson. Snowflakes. Digital Image. *Water.* Wikipedia. 1902. Public Domain.

15. USGS. The Water Cycle. Digital Image. *Water.* Wikipedia. 7. Sept. 2005.

16. Ash, Brian. More sunken curb at 7sigma Minneapolis, MN. Digital Image. *Rain Gardens.* Wikipedia. Public Domain.

17. Jones, Naia. Rain barrel at home in Seattle. Digital Image. *Seattle Rain Barrels.* Seattle Rain Barrels. Public Domain.

18. Gralo (Wikipedia Author). Nuclear Power Plant in Cattenom, France. Digital Image. *Nuclear Power Plants.* Wikipedia. 11. Mar. 2007. Creative Commons Attribution ShareAlike 3.0. See terms under Creative Commons Section.

19. Sandia National Laboratory. *Solar Energy.* Wikipedia. Public Domain.

CREATIVE COMMONS

Naia Jones is a professional civil engineer with ten years of experience in the industry. Her passion for living smarter with respect to the environment inspires her to write about the topic and to put smarter living concepts into practice. She launched the successful Seattle Rain Barrels company, a business providing affordable rain capturing devices for the average homeowner in the greater Seattle area. Naia currently lives in Honolulu, Hawaii with her family in a home powered by the sun.

www.ingramcontent.com/pod-product-compliance
Lightning Source LLC
Chambersburg PA
CBHW022126280326
41933CB00007B/558